BIOMONITORING OF TRACE AND TOXIC METALS IN HUMANS

Dr. Amit Singh Thakur

Kindle Direct Publishing

ISBN: 978-93-343-9592-1

CONTENTS

Biomonitoring of trace and toxic metals in Humans

Dr. Amit Singh Thakur
M.Sc. (Gold Medalist), Ph.D.
Assistant Professor, Department of Chemistry,
IPS Academy, Indore (M.P.) India

Publisher

Kindle Direct Publishing

INTRODUCTION

Biomonitoring, a vital field in environmental and public health science, involves the measurement and analysis of biological indicators within living organisms to assess exposure to various pollutants and chemicals. By studying these indicators, such as biomarkers or specific molecules, scientists gain insights into the presence and levels of contaminants present in the environment and their potential impacts on human health. Biomonitoring serves as a powerful tool to track trends in exposure, identify potential health risks, and evaluate the effectiveness of regulations and interventions. This field plays a crucial role in enhancing our understanding of the complex interactions between the environment and human health, ultimately guiding decisions to safeguard both the well-being of individuals and the ecosystems they inhabit.

BIOMONITORING: UNVEILING ENVIRONMENTAL HEALTH THROUGH LIVING ORGANISMS

In an era of rapidly advancing technology and increasing concern about the environment's impact on human health, the field of biomonitoring has emerged as a critical tool for understanding and addressing these complex interactions. Biomonitoring, a multidisciplinary scientific approach, involves the analysis of biological materials within living organisms to assess their exposure to various pollutants, chemicals, and environmental stressors. This innovative practice has the potential to revolutionize how we perceive and manage environmental health risks.

The Essence of Biomonitoring

At its core, biomonitoring provides a direct link between environmental contamination and its potential effects on human health. Instead of relying solely on environmental samples or theoretical models, biomonitoring examines real-time inter-

actions within organisms themselves. This is achieved by measuring specific biomarkers, which are indicators of exposure or biological responses, present within the organisms. These biomarkers might include metabolites, enzymes, or even DNA damage – all of which offer a unique perspective on the ongoing impact of environmental pollutants.

One of the most significant advantages of biomonitoring is its ability to capture nuanced data that traditional environmental sampling might miss. This technique can account for individual variability in susceptibility to toxins, as well as the cumulative effects of long-term exposure. As a result, biomonitoring can highlight health risks that might otherwise remain concealed.

Biomonitoring's applications are diverse and far-reaching. From urban environments to remote ecosystems, this practice has been utilized to study everything from heavy metal contamination in industrial areas to pesticide exposure in agricultural communities. Furthermore, biomonitoring can be employed to assess the effectiveness of regulatory measures, track changes in exposure levels over time, and even guide urban planning to minimize health risks.

While biomonitoring holds immense promise, it also presents challenges that must be addressed. Interpreting biomarker data requires a comprehensive understanding of the complex interactions between environmental factors, genetics, and individual behaviors. Moreover, ethical concerns related to privacy and informed consent must be navigated when using human participants in biomonitoring studies. In the future, biomonitoring could become even more sophisticated through integration with other technologies, such as wearable sensors and advanced imaging techniques. This could allow for real-time monitoring of exposure and health responses, enabling more precise interventions and personalized risk assessments.

Biomonitoring has the potential to transform the way we evaluate and respond to environmental health risks. By bridging the gap between environmental contamination and its effects on living organisms, this practice empowers policymakers, scientists, and communities to make informed decisions that safeguard both human health and the planet. As we continue to grapple with complex environmental challenges, biomonitoring stands as a beacon of hope, offering a clearer understanding of the intricate web linking our well-being to the world around us.

HISTORICAL BACKGROUND

The historical background and evolution of research on biomonitoring of human tissues and fluids is a journey that reflects our growing understanding of the complex interactions between environmental pollutants and human health. Here's a summarized timeline of key developments:

Early Observations (Ancient Times - 19th Century):

- Ancient civilizations observed the effects of environmental exposures on health, such as lead poisoning in Roman times due to lead pipes and wine storage containers.

- Industrialization during the 18th and 19th centuries led to increased awareness of occupational hazards, such as mercury exposure in hat-making factories ("mad hatter" syndrome).

Emergence of Toxicology (19th Century - Early 20th Century):

- The field of toxicology began to formalize, with researchers investigating the effects of various chemicals on health.

- Recognition of occupational diseases and the need for exposure monitoring led to the development of methods to quantify exposure to toxic substances.

Industrial Hygiene and Regulation (Mid 20th Century):

- As industrialization continued, occupational health and safety regulations were introduced.

- Monitoring methods for air and water quality expanded, but biomonitoring of human tissues and fluids was still in its infancy.

Advancements in Analytical Techniques (Late 20th Century):

- The mid to late 20th century saw significant advancements in analytical chemistry techniques, such as gas chromatography and mass spectrometry.

- These techniques allowed for the detection and quantification of a wide range of pollutants and their metabolites in biological samples.

Recognition of Chronic Effects (Late 20th Century - Early 21st Century):

- Researchers began to recognize the chronic and long-term health effects of exposure to low levels of pollutants over time.
- The concept of "body burden," the cumulative accumulation of pollutants in the body, gained attention.

Birth of Biomonitoring (Late 20th Century - Present):

- In the late 20th century, the term "biomonitoring" emerged to describe the practice of using biological samples to assess exposure to pollutants.

- The advent of advanced analytical techniques, like high-resolution mass spectrometry, allowed for the detection of even trace amounts of pollutants in human tissues and fluids.

Global Initiatives and Policies (21st Century):

- The 21st century brought increased recognition of the global nature of environmental pollution and the need for standardized biomonitoring approaches.

- National and international organizations started to establish guidelines and initiatives for biomonitoring efforts, aiming to understand population-level exposure trends.

Integration of Omics Technologies (21st Century):

- The integration of genomics, proteomics, metabolomics, and other "omics" technologies provided a deeper understanding of how pollutants affect molecular pathways and health outcomes.

Personalized and Exposome Research (21st Century):

- The concept of the "exposome," which considers the totality of exposures an individual experiences over their lifetime, gained prominence.

- Biomonitoring evolved to consider individual variability, genetics, and lifestyle factors, leading to personalized exposure assessments.

Advancements in Data Integration and Risk Communication (21st Century):

- Advances in data integration and computational methods allowed for the analysis of large datasets and the identification of complex exposure-response relationships.

- The importance of effective risk communication and public en-

gagement in biomonitoring efforts became evident.

In summary, the evolution of research on biomonitoring of human tissues and fluids has been marked by a journey from early observations of health effects to the development of sophisticated analytical techniques and the recognition of chronic exposure risks. This field continues to evolve with advancements in technology, growing awareness of global environmental challenges, and the integration of multiple disciplines. Today, biomonitoring stands as a crucial tool for understanding how environmental pollutants impact human health and informing strategies for minimizing exposure risks.

CURRENT TRENDS AND DEVELOPMENTS IN BIOMONITORING OF HUMAN TISSUES AND FLUIDS

Here are some current trends and developments that were prominent in the field of biomonitoring of human tissues and fluids here:

1. Advancements in Analytical Techniques: There has been a continuous development of advanced analytical techniques with higher sensitivity and precision. Mass spectrometry, including high-resolution mass spectrometry and tandem mass spectrometry (MS/MS), has become a cornerstone in biomonitoring. These techniques allow for the identification and quantification of a wide range of pollutants and biomarkers at very low concentrations.

2. Omics Technologies: The integration of omics technologies, such as genomics, transcriptomics, proteomics, and metabolomics, has enabled a more comprehensive understanding

of the biological responses to environmental exposures. These approaches provide insights into molecular pathways affected by pollutants and offer a broader perspective on the impact of exposures on health.

3. Non-Invasive Sampling Methods: Researchers are increasingly exploring non-invasive methods for sample collection, such as using saliva, exhaled breath, and even wearable devices. These methods offer convenience and may allow for more frequent monitoring of exposures without the need for invasive procedures.

4. Real-Time Monitoring: The development of sensors and wearable devices capable of real-time monitoring of biomarkers and pollutants is an emerging trend. This technology has the potential to provide immediate feedback on exposure levels and health status, allowing individuals to take proactive measures to minimize risks.

5. Integration of Data: The integration of biomonitoring data with other sources of information, such as environmental data and health records, is becoming more prevalent. This data fusion enables researchers to better understand the complex relationships between exposures, health outcomes, and individual susceptibilities.

6. Exposome Research: The exposome concept involves studying the totality of exposures an individual experiences over their lifetime. This approach goes beyond single pollutant analysis and considers the cumulative effects of various environmental factors, including pollutants, diet, lifestyle, and socioeconomic factors.

7. Global Collaborations and Initiatives: International collaborations and initiatives are on the rise to harmonize biomonitoring efforts and establish global databases of biomarker and exposure data. This allows for comparisons across different regions and populations, enhancing our understanding of worldwide environmental health trends.

8. Risk Communication and Education: There is an increasing emphasis on communicating biomonitoring results to the public, policymakers, and healthcare providers. Effective risk communication helps raise awareness about environmental health concerns and empowers individuals to make informed choices to reduce exposures.

9. Personalized Exposure Assessment: Advances in technology and data analysis are paving the way for personalized exposure assessments. This involves considering an individual's unique lifestyle, genetics, and environment to provide tailored

information about their specific exposure risks.

10. Ethical Considerations and Privacy: With the collection of more personal health data, ethical considerations and privacy protection have become paramount. Striking a balance between data utility and protecting individual privacy remains a challenge.

11. Regulatory Impact: Biomonitoring data continue to influence regulatory decisions and policies related to environmental protection and public health. As biomonitoring techniques improve, regulatory agencies may be more inclined to adopt biomonitoring data in setting exposure limits and guidelines.

CONCEPTS RELATED TO BIOMONITORING OF HUMAN TISSUES AND FLUIDS

Biomonitoring of human tissues and fluids involves the analysis of biological samples such as blood, urine, hair, and tissues to assess the presence and levels of various chemicals, pollutants, and biomarkers. This practice provides crucial insights into environmental exposure and its potential impacts on human health. Here's an overview of the main theories and concepts related to biomonitoring of human tissues and fluids:

1. Exposure Assessment: Exposure assessment is a fundamental concept in biomonitoring. It involves measuring the levels of environmental contaminants or their metabolites present in biological samples. This information helps quantify the extent of an individual's exposure to pollutants and assess any associated health risks.

2. Biomarkers: Biomarkers are measurable indicators of bio-

logical responses or changes resulting from exposure to environmental agents. They can be chemicals, proteins, enzymes, DNA damage, or other molecules that provide evidence of exposure, susceptibility, or early effects of exposure-related health outcomes.

3. Dose-Response Relationship: This concept examines the relationship between the amount of exposure to a pollutant and the biological response or health outcome. Dose-response relationships help establish safe exposure limits and predict health risks at different levels of exposure.

4. Threshold: The threshold refers to the point at which exposure to a pollutant begins to cause adverse health effects. Below this point, exposure may be considered safe or have no observable impact.

5. Half-life: The half-life of a substance in the body is the time it takes for half of the substance to be eliminated. Understanding the half-life of pollutants in human tissues and fluids helps assess the persistence of exposure.

6. Bioaccumulation and Biomagnification: Bioaccumulation occurs when pollutants accumulate in an organism over time due to exposure. Biomagnification refers to the increase in pollutant concentrations as they move up the food chain. Both con-

cepts are essential to understand how pollutants travel through ecosystems and impact human health.

7. Sampling and Analysis: Proper sampling and analysis techniques are crucial in biomonitoring. Samples need to be collected, stored, and processed carefully to avoid contamination and ensure accurate results. Analytical methods, such as gas chromatography, mass spectrometry, and atomic absorption spectroscopy, are used to quantify pollutant levels.

8. Reference Values and Baseline Data: Reference values, such as biological reference values (BRVs) and biological exposure indices (BEIs), provide benchmarks for normal pollutant levels in a population. Baseline data establish the typical levels of pollutants in a specific population, aiding in the identification of deviations from the norm.

9. Personal vs. Population Monitoring: Biomonitoring can be conducted on an individual level to assess personal exposure and health risks. Alternatively, population-level biomonitoring provides insights into broader trends and allows for the identification of vulnerable groups.

10. Temporal and Spatial Variation: Environmental exposures can vary over time and across geographic regions. Monitoring these variations helps to understand seasonal influences,

regional differences, and changes in exposure patterns.

11. Ethical and Privacy Considerations: Biomonitoring raises ethical concerns related to informed consent, privacy, and the sharing of individual exposure data. Balancing the benefits of knowledge gained from biomonitoring with participants' rights and privacy is a critical consideration.

12. Regulatory and Policy Implications: Biomonitoring data can influence regulatory decisions and policies related to environmental protection and public health. Governments and regulatory bodies can use this data to set exposure limits, establish guidelines, and enforce regulations.

Biomonitoring of human tissues and fluids plays a pivotal role in understanding the complex interactions between environmental contaminants and human health. The integration of exposure assessment, biomarkers, dose-response relationships, and ethical considerations provides a holistic view of the risks posed by environmental pollutants, guiding efforts to protect both individuals and the environment.

BIOMONITORING USING ATOMIC ABSORPTION SPECTROSCOPY:

In the pursuit of understanding the intricate relationship between environmental pollutants and human health, scientific innovation continually takes center stage. Among the array of analytical techniques available, atomic absorption spectroscopy (AAS) emerges as a powerful tool in the realm of biomonitoring, shedding light on the presence and impact of trace elements and heavy metals within living organisms. This article delves into the fascinating world of biomonitoring using AAS, exploring its principles, applications, and the insights it provides for safeguarding both ecosystems and human well-being.

Unveiling the Principles of Atomic Absorption Spectroscopy

Atomic absorption spectroscopy operates on the fundamental principle of light absorption. When light of a specific wave-

length passes through a sample containing atoms in a ground state, some of the light is absorbed by the atoms, causing them to transition to an excited state. By measuring the amount of absorbed light, researchers can quantify the concentration of the target element in the sample. This technique is particularly suited for analyzing trace metals and elements that have distinct absorption spectra.

Biomonitoring's Crucial Role in Environmental Health

The environment, teeming with a multitude of elements, both essential and toxic, significantly impacts human health. Biomonitoring using AAS provides a direct window into this complex relationship. By analyzing biological samples like blood, urine, hair, and tissues, scientists can detect the presence of elements like lead, mercury, cadmium, and arsenic – substances that, even in minute quantities, can have profound effects on health. This information is invaluable for assessing exposure levels, tracking trends over time, and formulating strategies to mitigate potential risks.

Applications in Real-World Contexts

The applications of biomonitoring through AAS are vast and

diverse. From monitoring industrial workers' exposure to heavy metals to assessing the impact of pollutants on aquatic organisms, this technique delivers insights into the effects of environmental contamination across various ecosystems. In regions with a history of mining, for example, AAS-enabled biomonitoring has been pivotal in identifying elevated levels of lead and other toxic metals, prompting targeted interventions to protect communities' well-being.

The Power of Precise Data

One of the most compelling features of AAS-based biomonitoring is its precision. Unlike traditional methods that rely on environmental samples, this approach provides data on the actual accumulation of elements within organisms. This accurate information is particularly relevant for vulnerable populations, such as children and pregnant women, who might be disproportionately affected by even low levels of certain pollutants.

Challenges and Advancements

While AAS has revolutionized biomonitoring, challenges persist. Sample collection and preservation, method validation, and accounting for confounding factors demand careful consideration. Nevertheless, ongoing advancements, such as the coupling of AAS with other analytical techniques, automation, and improved sensitivity, continue to enhance the accuracy and efficiency of this methodology.

Paving the Way Forward

Biomonitoring using atomic absorption spectroscopy exemplifies the symbiotic relationship between cutting-edge science and practical applications. As environmental concerns and health risks escalate, the insights gained from this technique become ever more critical. By harnessing the power of AAS, scientists, policymakers, and communities can collaborate to mitigate threats, preserve ecosystems, and ensure the health and prosperity of future generations. In the intricate dance between the environment and human well-being, AAS-enabled biomonitoring illuminates the path toward a harmonious coexistence.

HUMAN HAIR AS TOOL FOR MONITORING ENVIRONMENTAL POLLUTION

Human hair has emerged as a valuable and non-invasive tool for monitoring environmental pollution and assessing human exposure to various pollutants. This approach, known as "human hair biomonitoring," capitalizes on the fact that hair can accumulate a wide range of substances, including trace elements, heavy metals, organic compounds, and even drugs, reflecting both external exposures and internal processes. Here's how human hair serves as a tool for monitoring environmental pollution:

Accumulation of Contaminants: Hair grows from hair follicles that are supplied with blood vessels. As blood circulates through the body, it carries with it various substances, including pollutants. During hair growth, these substances are incorporated into the hair shaft, leading to their accumulation over time. This

accumulation pattern allows for the retrospective analysis of long-term exposure.

Non-Invasive Sampling: Collecting hair samples is non-invasive and relatively simple compared to other biomonitoring methods that might require blood or tissue samples. Hair can be easily collected from various parts of the body, such as the scalp or body hair, making it a convenient method for diverse populations and settings.

Temporal Information: Human hair grows at an average rate of about 1 cm per month, which means that different segments of the hair represent different time periods of exposure. This temporal information provides a timeline of exposure history, allowing researchers to distinguish short-term spikes in exposure from long-term trends.

Assessment of Cumulative Exposure: Hair can provide insights into cumulative exposure to pollutants over extended periods, which is particularly relevant for assessing the effects of chronic exposure. This is in contrast to blood or urine samples, which reflect more recent exposures.

Diverse Pollutants: Human hair can accumulate a wide range of pollutants, including heavy metals like lead, cadmium, and mercury, as well as organic compounds like pesticides, polychlorin-

ated biphenyls (PCBs), and drugs. This diversity allows for monitoring exposure to various types of contaminants.

Spatial Variation: Hair samples can capture spatial variation in exposure within a population, making them useful for assessing regional differences in pollutant exposure.

Ethnic and Genetic Considerations: Human hair biomonitoring takes into account factors such as hair color, texture, and ethnic differences, which can affect the rate of pollutant accumulation. This consideration enhances the accuracy of exposure assessments across diverse populations.

Challenges and Considerations: Despite its advantages, human hair biomonitoring has some limitations. External contamination, such as through shampoos or hair treatments, can influence results. Interpretation can also be complex due to factors like individual hair growth rates and the potential for pollutants to bind differently to different hair types.

In conclusion, human hair serves as a valuable tool for monitoring environmental pollution and assessing human exposure to a wide array of pollutants. Its non-invasive nature, ability to provide temporal information, and capacity to reflect cumulative exposure make it an attractive medium for biomonitor-

ing efforts, contributing to our understanding of environmental health risks and informing public health policies.

HUMAN FINGERNAILS AS TOOL FOR MONITORING ENVIRONMENTAL POLLUTION

Human fingernails have gained recognition as a unique and practical tool for monitoring environmental pollution and assessing the presence of various contaminants in the human body. This approach, often referred to as "human fingernail biomonitoring," offers distinct advantages and insights into an individual's exposure to pollutants and chemicals. Here's how human fingernails serve as a tool for monitoring environmental pollution:

Slow Growth and Accumulation: Similar to human hair, fingernails grow slowly over time, accumulating substances from the environment as they grow. This accumulation occurs as blood circulates nutrients and chemicals to the nail matrix, where they become part of the growing nail. This slow growth rate allows for the integration of exposure data over a period of

months, providing a retrospective view of exposure.

Non-Invasive Sampling: Collecting fingernail samples is a non-invasive and straightforward procedure. Unlike collecting blood or tissue samples, fingernail collection does not require needles or complex procedures. It can be done using simple nail clippings, making it more comfortable and accessible for participants.

Longitudinal Exposure Assessment: Fingernails can provide a cross-sectional assessment of exposure along the length of the nail, allowing researchers to analyze exposure changes over time. This can be especially useful for identifying shifts in exposure patterns or changes in pollutant sources.

Diverse Contaminants: Human fingernails can accumulate a variety of contaminants, including heavy metals, organic pollutants, and trace elements. This diversity of substances provides a comprehensive picture of an individual's exposure to different types of pollutants.

Individual Variation: Fingernail biomonitoring takes into account factors such as nail growth rates and individual behaviors that can influence the accumulation of pollutants. This consideration allows for more accurate exposure assessments tailored to each individual.

Relationship to Health Outcomes: Research suggests that certain elements and pollutants present in fingernails can be associated with specific health outcomes. For example, elevated levels of certain heavy metals in fingernails may indicate an increased risk of health issues related to metal toxicity.

Environmental Exposure Tracking: Fingernail biomonitoring is especially useful for tracking exposure to contaminants that are not rapidly metabolized or eliminated from the body. It can provide insights into long-term exposure trends, complementing other biomonitoring methods that focus on more immediate exposures.

Challenges and Considerations: While fingernail biomonitoring offers unique advantages, there are some challenges to consider. External contamination from nail treatments or contact with contaminated surfaces can affect results. Standardization of sampling procedures and analytical methods is important to ensure consistency across studies.

Human fingernails are a valuable tool for monitoring environmental pollution and assessing individual exposure to various pollutants. Their slow growth, non-invasive sampling, and ability to accumulate diverse contaminants make them an attractive medium for biomonitoring efforts. Fingernail biomonitoring contributes to our understanding of long-term exposure

trends, aids in identifying potential health risks, and informs strategies for mitigating environmental pollutants' impact on human health.

HUMAN URINE AS TOOL FOR MONITORING ENVIRONMENTAL POLLUTION

Human urine is a highly informative and widely used tool for monitoring environmental pollution and assessing the presence of various contaminants in the body. This practice, known as "human urine biomonitoring," offers insights into an individual's exposure to pollutants, as well as the potential health effects of these exposures. Here's how human urine serves as a tool for monitoring environmental pollution:

Dynamic Reflection of Recent Exposure: Urine is a dynamic fluid that rapidly reflects recent exposures to pollutants. Many chemicals and their metabolites are eliminated from the body through urine shortly after exposure. This makes urine biomonitoring particularly effective for assessing short-term exposure to pollutants.

Non-Invasive Collection: Collecting urine samples is non-invasive and relatively straightforward. Participants provide a urine sample, which can be conveniently collected in healthcare set-

tings or even in participants' homes. This ease of collection facilitates large-scale studies and enhances participant compliance.

Quantitative Analysis: Urine biomonitoring allows for quantitative analysis of both the concentration of pollutants and their metabolites. This information can be used to assess exposure levels, track trends over time, and compare with established reference values or guidelines.

Wide Range of Contaminants: Urine can contain a wide array of substances, including heavy metals, pesticides, organic pollutants, and pharmaceuticals. This diversity of pollutants reflects exposures from various sources and provides a comprehensive view of an individual's environmental contact.

Metabolism and Susceptibility: Some pollutants are metabolized by the body before being excreted in urine. By analyzing urinary metabolites, researchers can gain insights into the metabolic pathways and potential health effects of pollutants. Additionally, individual variations in metabolism can be taken into account.

Identifying Biomarkers: Biomarkers are specific substances or compounds present in urine that indicate exposure to particular pollutants. Researchers can identify and validate these biomarkers to create reliable indicators of exposure, even at low concentrations.

Population Health Insights: Urine biomonitoring data can provide insights into population-level exposure patterns and trends. This information can help inform public health policies and interventions to reduce exposure to specific contaminants.

Challenges and Considerations: While urine biomonitoring has several advantages, there are also challenges to consider. Urine concentrations of pollutants can vary depending on factors such as hydration status, individual differences in metabolism, and timing of sample collection. Proper standardization and consideration of these factors are crucial for accurate interpretation of results.

Ethical and Privacy Considerations: Handling personal bodily fluids like urine raises ethical considerations related to informed consent, privacy protection, and the responsible use of data for research purposes.

In conclusion, human urine is a powerful tool for monitoring environmental pollution and assessing human exposure to a wide range of contaminants. Its ability to rapidly reflect recent exposure, non-invasive collection, quantitative analysis, and potential for identifying biomarkers make it a valuable medium for biomonitoring efforts. Urine biomonitoring contributes to our understanding of short-term exposure trends, aids in identifying potential health risks, and informs strategies for reducing environmental pollutants' impact on human health.

HUMAN BLOOD AS TOOL FOR MONITORING ENVIRONMENTAL POLLUTION

Human blood serves as a vital and comprehensive tool for monitoring environmental pollution and assessing the presence of various contaminants in the body. "Human blood biomonitoring" is a practice that involves analyzing blood samples to gain insights into an individual's exposure to pollutants and potential health effects. Here's how human blood functions as a tool for monitoring environmental pollution:

Direct Reflection of Exposure: Blood is a direct route through which many chemicals and pollutants enter the body. It transports these substances to various organs and tissues, providing a real-time reflection of recent exposures. Blood biomonitoring captures both short-term and ongoing exposures to pollutants.

Transport of Pollutants and Metabolites: Blood carries pollu-

tants from the site of exposure to various parts of the body, including organs responsible for metabolism and excretion. It also carries metabolites, which are products resulting from the breakdown of pollutants by the body's processes. Analyzing these metabolites offers insights into how the body processes pollutants.

Quantitative Analysis: Blood biomonitoring allows for quantitative analysis of the concentration of pollutants and their metabolites. This data can be used to assess exposure levels, track trends over time, and compare with established reference values or guidelines.

Assessment of Internal Exposure: Blood represents internal exposure to pollutants, as it circulates throughout the body, reaching various tissues and organs. This is in contrast to external exposure measurements that might only capture surface contact.

Variety of Contaminants: Blood can contain a wide range of pollutants, including heavy metals, organic chemicals, pesticides, and pharmaceuticals. This diversity of pollutants reflects exposures from different sources and provides a comprehensive view of an individual's exposure profile.

Identification of Biomarkers: Blood can contain specific biomarkers—biological indicators—that show exposure to particu-

lar pollutants. Researchers can identify and validate these biomarkers to create reliable indicators of exposure, even at low concentrations.

Association with Health Outcomes: Blood biomonitoring data can be correlated with health outcomes to understand potential risks. Elevated levels of certain pollutants in the blood may indicate an increased risk of adverse health effects.

Clinical and Regulatory Relevance: Blood biomonitoring data can have clinical relevance, helping healthcare providers identify exposure-related health issues and guiding patient management. Additionally, these data can influence regulatory decisions and policies aimed at reducing environmental exposures.

Challenges and Considerations: Blood biomonitoring has several advantages, but it also presents challenges. Blood collection involves invasive procedures, which might impact participant compliance. Ethical considerations, informed consent, and privacy protection are crucial aspects when collecting and using blood samples for research.

Temporal Variation: Blood concentrations of pollutants can vary over time due to factors like meal consumption, physical activity, and time of day. Proper consideration of these factors is essential for accurate interpretation of blood biomonitoring

results.

Human blood is a powerful tool for monitoring environmental pollution and assessing individual exposure to a wide range of contaminants. Its direct reflection of exposure, internal exposure assessment, quantitative analysis, and potential for identifying biomarkers make it a valuable medium for biomonitoring efforts. Blood biomonitoring contributes to our understanding of both short-term and ongoing exposure trends, aids in identifying potential health risks, and informs strategies for reducing environmental pollutants' impact on human health.

ENVIRONMENTAL EXPOSURE ASSESSMENT IN INDIAN POPULATIONS

Environmental exposure assessment plays a pivotal role in understanding the interactions between the environment and human health. In the context of Indian populations, a diverse and rapidly evolving landscape presents unique challenges and opportunities for studying the impact of environmental pollutants on public health. India's vast geography, varied socio-economic conditions, and complex mix of urban and rural settings create a rich tapestry for environmental exposure assessment.

India's environmental exposure assessment efforts encompass a wide spectrum of pollutants, ranging from heavy metals, organic chemicals, and pesticides to air pollutants and emerging contaminants. Researchers and public health agencies have embarked on numerous studies to evaluate the levels of these

contaminants in various environmental media, as well as their uptake in human tissues and fluids.

The diverse geography and industrial activity of India contribute to varying exposure scenarios. Urban centers grapple with issues such as air pollution from vehicular emissions, industrial discharges, and indoor pollution. Meanwhile, rural areas may face challenges related to pesticide exposure, water contamination, and traditional practices that introduce pollutants into the environment.

Government agencies, research institutions, and non-governmental organizations in India are actively engaged in generating exposure data to inform policy decisions and public health interventions. Studies have examined pollutants in drinking water, agricultural produce, air quality in urban centers, and occupational exposure among workers in various industries. Biomonitoring efforts using human tissues and fluids have shed light on the internal burdens of contaminants, offering insights into the level of exposure among different population groups.

One of the challenges faced in environmental exposure assessment in India is the need for harmonized methodologies and standardized sampling protocols across regions. Additionally, bridging the gap between research findings and actionable policies remains a priority. The information gleaned from environmental exposure assessments serves as a vital foundation for formulating regulations, designing interventions, and raising awareness among the public. As India navigates its journey towards sustainable development, robust exposure assessment will continue to be a cornerstone in safeguarding public health, preserving the environment, and achieving a harmonious balance between progress and well-being.

Human biomonitoring studies in India provide invaluable insights into the complex relationship between environmental pollutants and public health. These studies serve as a critical bridge between exposure assessment, health outcomes, and policy formulation. With its diverse population, varied environmental contexts, and emerging health challenges, India stands as a unique and dynamic setting for conducting human biomonitoring research.

Biomonitoring endeavors in India encompass a broad spectrum of pollutants, including heavy metals, persistent organic pollutants, endocrine-disrupting chemicals, and contaminants arising from urbanization and industrialization. These studies often involve the collection and analysis of biological samples such as blood, urine, hair, and breast milk to quantify the presence of pollutants and assess their potential health impacts.

Through biomonitoring, researchers gain a deeper understand-

ing of exposure patterns across different regions and demographic groups. Urban areas contend with pollutants from vehicular emissions, industrial processes, and indoor sources, while rural populations face challenges like pesticide exposure, contaminated water, and traditional practices that may introduce pollutants.

Several governmental bodies, research institutions, and collaborative initiatives in India have undertaken biomonitoring studies to inform policies and public health interventions. These studies contribute to setting exposure standards, identifying vulnerable populations, and raising awareness about the importance of minimizing exposure to environmental contaminants.

Human biomonitoring studies also play a pivotal role in documenting the changing landscape of exposure. They provide a longitudinal view of exposure trends, helping policymakers anticipate emerging environmental health risks and adjust strategies accordingly.

While progress has been made, challenges remain. Biomonitoring studies in India need to address issues such as participant recruitment, ethical considerations, data interpretation, and the need for capacity-building across institutions. Moreover, collaboration between researchers, policymakers, healthcare profes-

sionals, and the public is essential to translate research findings into meaningful actions.

In a rapidly evolving world, human biomonitoring studies continue to unveil the intricate connections between environmental quality and human well-being. By fostering a holistic understanding of the exposures faced by Indian populations, these studies serve as catalysts for informed decision-making, improved regulations, and the protection of public health in the face of a changing environment.

BIOMONITORING OF ORGANIC POLLUTANTS IN INDIA

Biomonitoring studies focused on organic pollutants in India contribute significantly to our understanding of the impact of these contaminants on human health and the environment. Organic pollutants encompass a wide range of compounds, including pesticides, industrial chemicals, and other synthetic substances that have become integral to modern life. As India grapples with the challenges of rapid industrialization, urbanization, and agricultural intensification, organic pollutants biomonitoring plays a crucial role in safeguarding the well-being of its population.

In both urban and rural settings, organic pollutants biomonitoring sheds light on exposures that arise from various sources, such as the use of pesticides in agriculture, the release of industrial chemicals, and the persistence of pollutants in the environment. Studies often involve the analysis of biological samples, such as blood, urine, and breast milk, to assess the presence and

levels of these contaminants.

India's diverse geography and agricultural practices contribute to a complex interplay of organic pollutants. Pesticide residues find their way into food and water sources, raising concerns about chronic exposure and potential health effects. Additionally, urban centers contend with pollutants from industrial processes and vehicular emissions, which can have far-reaching implications for respiratory health and overall well-being.

Biomonitoring efforts in India go beyond identifying the presence of pollutants; they aim to uncover links between exposure and adverse health outcomes. These studies contribute to evidence-based decision-making, helping authorities set exposure limits, formulate regulations, and develop strategies to mitigate risks.

Collaborative initiatives involving government agencies, research institutions, and non-governmental organizations play a crucial role in advancing organic pollutants biomonitoring in India. By fostering partnerships, sharing data, and building capacity, these efforts contribute to a comprehensive understanding of the complex challenges posed by organic pollutants.

However, challenges persist, including the need for standardized methodologies, capacity building, and long-term monitor-

ing efforts. Ethical considerations, community engagement, and effective risk communication are essential aspects that need to be integrated into biomonitoring initiatives.

Organic pollutants biomonitoring in India represents a proactive stance towards environmental and public health concerns. By uncovering the intricate relationship between human exposure and pollutants, these studies are a driving force behind informed policies, healthier communities, and a sustainable future.

BIOMONITORING OF PESTICIDES IN INDIA

Pesticides play a crucial role in modern agriculture, protecting crops and ensuring food security. However, the widespread use of these chemicals also raises concerns about potential health risks for both agricultural workers and the general population. Pesticide exposure biomonitoring in India is a vital endeavor that sheds light on the extent of exposure, potential health effects, and the need for sustainable agricultural practices.

India's vast agricultural landscape is a mix of smallholder farming and large-scale cultivation. Pesticides are used to combat pests and enhance crop yield, but improper handling, lack of protective measures, and inadequate training can lead to unintended exposure among farmers and farm laborers.

Biomonitoring studies in India focus on assessing the presence and levels of pesticide residues in biological samples, such as blood, urine, and hair. These studies help determine the extent of exposure and the types of pesticides to which individuals are being exposed. This information is crucial for understanding

health risks and informing policies that can protect vulnerable populations.

Rural areas, where the majority of India's population resides, often face higher pesticide exposure due to agricultural activities. Pesticide residues can find their way into water sources, affecting drinking water and food crops. These exposures can have far-reaching health implications, including acute poisoning, chronic health effects, and potential impacts on future generations.

Efforts to monitor pesticide exposure involve collaboration among government agencies, research institutions, and non-governmental organizations. These initiatives aim to raise awareness among farmers, promote safer pesticide use practices, and advocate for the adoption of integrated pest management strategies that minimize reliance on chemical pesticides.

Challenges in pesticide exposure biomonitoring include the need for accurate methodologies, ensuring representative sampling, and addressing ethical considerations in working with vulnerable populations. Language barriers, cultural practices, and accessibility to healthcare facilities are factors that require careful consideration in these studies.

Pesticide exposure biomonitoring in India is a proactive step

towards creating a healthier and more sustainable agricultural sector. By illuminating the risks associated with pesticide use and advocating for safer practices, these studies contribute to the well-being of both agricultural workers and consumers. They also underscore the importance of striking a balance between agricultural productivity and environmental and public health concerns.

METALS IN HUMAN TISSUES IN INDIA: UNDERSTANDING EXPOSURES AND IMPLICATIONS

Metals, both essential and toxic, are an integral part of our environment. While essential metals like iron and zinc are vital for health, heavy metals such as lead, mercury, cadmium, and arsenic can pose significant health risks when present in excessive amounts. In India, a country with diverse environmental and industrial contexts, the study of metals in human tissues offers a profound insight into the interactions between environmental quality and public health.

Exposure Pathways and Vulnerable Groups:

Metals can enter the human body through various exposure pathways, including contaminated air, water, soil, and food. In India, rapid urbanization, industrialization, and mining activ-

ities contribute to metal pollution in urban and rural areas alike. Industrial emissions, vehicular pollution, and improper waste disposal are sources of heavy metal contamination. Vulnerable groups such as children, pregnant women, and marginalized communities are at higher risk due to physiological factors, living conditions, and dietary habits.

Biomonitoring Studies:

Biomonitoring studies involving the analysis of metals in human tissues—such as blood, urine, hair, and nails—provide a comprehensive view of exposures. These studies measure both essential and toxic metals, allowing researchers to assess potential imbalances and health implications. For instance, lead exposure, often linked to lead-acid battery recycling and lead-based paints, has been a significant concern in Indian cities. Arsenic contamination in groundwater, especially in certain regions, is another pressing issue due to its association with chronic health effects.

Implications for Health:

Excessive exposure to toxic metals can lead to a range of health effects, including neurological disorders, developmental delays, cardiovascular diseases, and cancers. Essential metals, when imbalanced, can disrupt normal physiological functions. In India, metal-related health concerns range from lead poisoning in children to arsenic-induced skin lesions and respiratory problems.

Government Initiatives and Policies:

The Indian government has taken steps to address metal pollution and its impact on human health. Guidelines and regulations have been established for permissible limits of metals in drinking water, air quality standards, and industrial emissions. However, enforcing these regulations and ensuring compliance across diverse settings remain challenges.

Community Awareness and Interventions:

Biomonitoring studies not only provide insights into exposures but also serve as tools for community awareness and education. Local communities are empowered when they understand the risks associated with metal exposures and adopt preventive measures. Efforts to promote safer practices, such as using clean water sources and proper waste disposal, play a vital role in reducing metal exposures.

Research Collaboration and Capacity Building:

Collaboration between governmental bodies, research institutions, non-governmental organizations, and international agencies is essential for comprehensive research, data sharing, and capacity building. This collaboration strengthens India's ability to assess exposures, evaluate risks, and devise effective interventions.

In conclusion, the study of metals in human tissues in India offers a window into the complex interplay between environmental pollutants and human health. By unraveling exposure patterns, health effects, and societal implications, these studies contribute to informed decision-making, policy formulation, and community well-being. They underscore the importance of a holistic approach that addresses the multifaceted challenges posed by metal pollution in a country where environmental and public health are closely intertwined.

Heavy Metals Biomonitoring in India: Assessing Risks and Protecting Health**

Heavy metals, known for their toxicity and persistence, are among the most concerning environmental pollutants globally. In India, a country marked by diverse landscapes and rapid industrialization, the biomonitoring of heavy metals in human tissues provides crucial insights into exposures, potential health impacts, and the need for effective regulations and interventions.

Heavy Metal Contaminants and Sources:

Heavy metals such as lead, mercury, cadmium, arsenic, and chromium originate from both natural processes and human activities. Anthropogenic sources include industrial emissions, mining activities, vehicular exhaust, and improper waste disposal. The unique mix of urban and rural environments in India amplifies the complexity of heavy metal exposure pathways.

Biomonitoring Studies and Methods:

Biomonitoring studies involve the collection and analysis of biological samples—blood, urine, hair, nails—to quantify heavy metal concentrations. These studies help researchers understand exposure patterns, regional variations, and the potential health implications. Sophisticated analytical techniques like atomic absorption spectroscopy and mass spectrometry enable accurate measurements even at trace levels.

Health Implications:

Heavy metals have a range of adverse health effects, from immediate symptoms to long-term chronic diseases. Lead exposure is associated with developmental delays in children, while arsenic exposure in groundwater can lead to skin lesions and increased cancer risks. Mercury can impact the nervous system, and cadmium exposure is linked to kidney damage and respiratory issues.

Vulnerable Populations and Health Disparities:

Certain populations, including children, pregnant women, and occupational groups, are particularly vulnerable to heavy metal exposures due to their physiology, behaviors, or work environments. Socioeconomic factors can exacerbate these vulnerabilities, leading to health disparities.

Government Regulations and Policies:

The Indian government has set permissible limits for heavy metal concentrations in drinking water, air, and soil. Regulatory bodies like the Central Pollution Control Board (CPCB) and the Ministry of Environment, Forest and Climate Change (MoEFCC) monitor and enforce these standards. However, challenges remain in effective implementation and enforcement, particularly in resource-constrained regions.

Community Engagement and Education:

Biomonitoring studies contribute to community awareness by providing tangible evidence of heavy metal exposures. Educating communities about potential risks and preventive measures empowers them to take steps to reduce exposures. NGOs, healthcare workers, and local leaders play pivotal roles in disseminating information.

Research Collaborations and Future Directions:

Collaborations between governmental agencies, research institutions, and international organizations are vital for advancing heavy metal biomonitoring research. Capacity building, data sharing, and standardized methodologies enhance the quality and impact of these studies.

Heavy metals biomonitoring in India plays a crucial role in understanding the complex relationship between environmental pollutants and public health. These studies provide evidence to inform policies, regulations, and interventions that can mitigate the risks posed by heavy metal exposures. By safeguarding human health and promoting sustainable practices, heavy metals biomonitoring contributes to the well-being of India's diverse population and the preservation of its environment.

SIGNIFICANCE OF HAIR AND NAIL OVER BLOOD AND URINE IN STUDYING ENVIRONMENTAL EXPOSURE

When it comes to studying environmental exposure to contaminants, biological samples like blood and urine have traditionally been the go-to choices due to their direct reflection of recent exposures. However, hair and nail samples are increasingly gaining significance in biomonitoring studies for several compelling reasons. These keratinized tissues offer unique advantages that contribute to a more comprehensive understanding of long-term exposure trends, cumulative burdens, and potential health implications.

Accumulation and Long-Term Exposure:

Hair and nails grow relatively slowly compared to the rapid turnover of blood and urine. This slow growth rate allows hair

and nail samples to accumulate contaminants over extended periods, offering insights into chronic and long-term exposures. While blood and urine provide snapshots of recent exposures, hair and nails provide a retrospective view of an individual's exposure history.

Temporal Information:

Hair and nails provide a chronological record of exposure due to their growth patterns. Different segments of hair and nails represent distinct time periods, allowing researchers to analyze exposures over weeks, months, or even years. This temporal information is invaluable for identifying exposure trends and distinguishing between short-term spikes and continuous exposure.

Cumulative Exposure Assessment:

Hair and nails act as repositories, accumulating contaminants as they grow. This accumulation pattern allows for the assessment of cumulative exposure over time. In contrast, blood and urine samples primarily reflect recent exposures, making them more suitable for capturing immediate or short-term exposure events.

Non-Invasive Sampling:

Collecting hair and nail samples is non-invasive and relatively straightforward compared to blood and urine collection. This ease of sampling enhances participant compliance and makes it feasible for larger-scale studies, especially in diverse or remote populations.

Individual Variation:

Hair and nail biomonitoring takes into account individual variations such as metabolic rates, hydration status, and genetic factors that might affect the accumulation and excretion of contaminants. This consideration enhances the accuracy of exposure assessments across diverse populations.

Identifying Chronic Exposures:

Some contaminants are slowly metabolized and excreted, making their detection challenging in blood and urine samples. Hair and nail analysis can provide insights into chronic exposures to these persistent contaminants, even when blood and urine levels might no longer reflect the exposure.

Monitoring Regional and Occupational Exposures:

Hair and nails can capture spatial variation in exposure within a population, making them useful for assessing regional differences in pollutant exposure. They are also particularly relevant for monitoring occupational exposures to specific contaminants.

Challenges and Considerations:

While hair and nail biomonitoring offer substantial advantages, challenges like external contamination, variations in growth rates, and the potential for contaminant binding should be considered during analysis and interpretation.

The significance of hair and nail samples in studying environmental exposure lies in their ability to provide a retrospective, cumulative, and long-term view of an individual's exposure history. By complementing the information gained from blood and urine samples, hair and nails contribute to a more comprehensive understanding of exposure trends, allowing researchers to better assess the potential health risks associated with chronic exposures to environmental contaminants.

INDUCTIVELY COUPLED PLASMA MASS SPECTROMETRY (ICP-MS) IN HUMAN BIOMONITORING: EXPLORING EXPOSURE TO TRACE ELEMENTS AND METALS

Inductively Coupled Plasma Mass Spectrometry (ICP-MS) is a powerful analytical technique that plays a pivotal role in human biomonitoring studies by enabling the accurate and sensitive quantification of trace elements and metals in biological samples. It is a versatile technique that offers high sensitivity, precision, and the ability to analyze a wide range of elements simultaneously. ICP-MS is particularly valuable for assessing exposure to heavy metals and trace elements, which can have both beneficial and toxic effects on human health.

Principle of ICP-MS:

ICP-MS operates based on the principles of mass spectrometry and utilizes an inductively coupled plasma as the ionization source. In this technique, a sample is first atomized and ionized by the high-temperature plasma. The resulting ions are then separated based on their mass-to-charge ratios using a mass analyzer, allowing for the quantification of different elements present in the sample.

Application in Human Biomonitoring:

ICP-MS is widely used in human biomonitoring studies to measure the concentrations of various trace elements and metals in biological samples such as blood, urine, hair, and nails. These elements include essential nutrients like zinc and selenium, as well as potentially toxic elements such as lead, mercury, cadmium, and arsenic. The technique's high sensitivity allows detection even at extremely low concentrations, making it suitable for quantifying trace levels of contaminants.

Sample Preparation:

Prior to analysis, biological samples need to be properly pre-

pared to ensure accurate results. This may involve digestion to break down organic matter, dilution to bring the sample within the instrument's working range, and the addition of internal standards for quantification accuracy and correction of instrumental variations.

Advantages of ICP-MS in Human Biomonitoring:

1. Sensitivity and Detection Limits: ICP-MS can detect elements at extremely low concentrations, often down to parts per billion (ppb) or even parts per trillion (ppt), making it suitable for assessing exposure to trace elements and metals.

2. Elemental Range: ICP-MS can analyze a wide range of elements across the periodic table, from light elements to heavy metals, allowing for comprehensive biomonitoring.

3. High Throughput: ICP-MS is a high-throughput technique, capable of analyzing multiple elements simultaneously in a single run, which is particularly advantageous for large-scale biomonitoring studies.

4. Quantitative Accuracy: The use of internal standards and calibration curves ensures accurate quantification, even in complex biological matrices.

5. Isotope Information: ICP-MS can provide isotopic information, allowing researchers to distinguish between different sources of an element and assess its bioavailability and metabolism.

Challenges and Considerations:

Despite its many advantages, ICP-MS has some challenges. Sample preparation can be labor-intensive, and there's a potential for contamination during the process. Background interferences can also affect accuracy, requiring careful selection of analytical conditions and use of interference correction techniques.

Conclusion:

ICP-MS has revolutionized the field of human biomonitoring by allowing researchers to accurately measure trace elements and metals in biological samples. Its sensitivity, versatility, and ability to quantify multiple elements make it an essential tool for assessing exposure to both essential nutrients and toxic contaminants, contributing to our understanding of how environmental exposures impact human health.

NEUTRON ACTIVATION ANALYSIS (NAA) IN HUMAN BIOMONITORING: UNVEILING ELEMENTAL FINGERPRINTS OF ENVIRONMENTAL EXPOSURE

Neutron Activation Analysis (NAA) is a sophisticated analytical technique that plays a significant role in human biomonitoring studies, providing insights into the presence and concentrations of trace elements and other substances in biological samples. NAA utilizes the unique property of certain isotopes to become radioactive upon neutron capture, allowing for the detection and quantification of elements present at extremely low levels. This technique has proven invaluable in assessing exposure to various elements and understanding their impact on human health.

Principle of NAA:

NAA involves the irradiation of a sample with neutrons in a nuclear reactor or accelerator. When stable isotopes in the sample capture neutrons, they become radioactive and emit characteristic gamma rays. These gamma rays are specific to each element and serve as a fingerprint for its presence and concentration in the sample. By measuring the emitted gamma rays, scientists can identify and quantify the elements.

Application in Human Biomonitoring:

NAA is particularly useful in human biomonitoring studies due to its sensitivity and ability to detect multiple elements simultaneously. It has been employed to analyze various biological samples such as blood, urine, hair, nails, and even bones. NAA helps researchers understand exposure to essential elements like calcium, magnesium, and zinc, as well as potentially toxic elements such as arsenic, mercury, and cadmium.

Advantages of NAA in Human Biomonitoring:

1. Sensitivity: NAA can detect elements at extremely low concentrations, even in the parts per billion (ppb) to parts per trillion (ppt) range, making it suitable for studying trace elements.

2. Multi-Element Analysis: NAA allows for the simultaneous analysis of multiple elements in a single measurement, providing a comprehensive view of the elemental composition of a sample.

3. Non-Destructive: NAA is a non-destructive technique, meaning that after analysis, the sample remains intact and can be further studied using other methods.

4. Elemental Identification: The characteristic gamma rays emitted during radioactive decay provide unique signatures for each element, allowing for accurate identification.

5. Minimal Sample Preparation: NAA requires minimal sample preparation, reducing the potential for contamination and preserving the integrity of the sample.

Challenges and Considerations:

While NAA offers numerous advantages, it also comes with certain challenges. Access to a nuclear reactor or accelerator for

sample irradiation is required, which limits the availability of the technique. Additionally, the analysis of gamma rays requires specialized equipment and expertise.

Conclusion:

Neutron Activation Analysis is a powerful tool in human bio-monitoring, enabling the accurate quantification of trace elements and substances in biological samples. Its ability to provide multi-elemental information at low concentrations contributes significantly to our understanding of environmental exposures and their potential impact on human health. NAA continues to be a valuable asset in research efforts aimed at uncovering the intricate relationships between human biology and the environment.

ACCUMULATION OF TOXIC COMPOUNDS IN HUMAN HAIR AND NAIL

The accumulation of toxic compounds in human hair and nails is a complex phenomenon influenced by several theories and concepts. Understanding these theories helps elucidate how these biomatrices can serve as valuable indicators of long-term exposure to environmental pollutants. Here's an overview of the main theories and concepts related to the accumulation of toxic compounds in human hair and nails:

Bioaccumulation and Biomagnification:

Bioaccumulation refers to the gradual accumulation of substances, including toxic compounds, in living organisms over time. Biomagnification is an associated concept where the con-

centration of certain compounds increases as one moves up the food chain. In the context of human hair and nails, these theories suggest that as individuals are exposed to pollutants from various sources (e.g., air, water, diet), these pollutants may accumulate in hair and nails due to their slow growth rates and the potential for binding and storage of certain substances.

Keratin as a Sink for Contaminants

Hair and nails are primarily composed of a protein called keratin, which contains sulfur-rich amino acids that can bind to metal ions and other contaminants. This binding capacity makes keratin an efficient sink for a variety of substances, leading to the accumulation of toxic compounds within hair and nail structures.

Sequential Deposition

The growth of hair and nails occurs in distinct phases, resulting in the sequential deposition of new material. This sequential growth can potentially result in layers representing different time periods, allowing for retrospective analysis of exposures over time. This concept is particularly useful for interpreting biomonitoring data obtained from hair and nail samples.

Long Turnover Times

Hair and nails have longer turnover times compared to other biomatrices like blood and urine. While blood and urine reflect recent exposures, hair and nails can capture chronic and cumulative exposures. The slow growth rates of hair and nails contribute to the long-term integration of toxic compounds.

Concentration Gradients

Hair and nails can accumulate contaminants at varying concentrations along their lengths. This concept is known as concentration gradients, and it occurs due to changes in exposure levels or metabolic activity during different growth phases. Analyzing different segments of hair or nails can provide insights into exposure trends over time.

External vs. Internal Contamination

Hair and nails can be externally contaminated by substances from the environment, such as dust or pollutants present in water. Distinguishing between external and internal contamination is crucial for accurate interpretation. Analyzing specific segments of hair or nails can help differentiate between these types of contamination.

Elemental Analysis Techniques

Advanced analytical techniques like Inductively Coupled Plasma Mass Spectrometry (ICP-MS) and Neutron Activation Analysis (NAA) enable the quantification of trace elements and metals in hair and nail samples. These techniques contribute to a deeper understanding of the accumulation patterns of toxic compounds.

The accumulation of toxic compounds in human hair and nails is influenced by concepts such as bioaccumulation, biomagnification, keratin binding, sequential deposition, slow turnover rates, and concentration gradients. These theories collectively demonstrate the value of using hair and nail samples as archives of long-term exposure to environmental pollutants, providing insights into historical exposures and potential health risks.

HUMAN EXPOSURE TO TRACE AND TOXIC ELEMENTS

Human exposure to trace and toxic elements is a complex topic influenced by various theories and concepts. These theories help us understand how individuals come into contact with these elements, how they enter and interact with the body, and the potential health implications. Here's an overview of the main theories and concepts related to human exposure to trace and toxic elements:

1. Routes of Exposure: Exposure to trace and toxic elements can occur through various routes, including inhalation (breathing contaminated air), ingestion (consuming contaminated food, water, or beverages), and dermal contact (skin contact with contaminated surfaces). These routes determine how elements enter the body and can influence the degree of exposure.

2. Source-to-Dose Relationship: This concept emphasizes the link between the source of elements (natural or anthropogenic) and the subsequent dose received by an individual. Understand-

ing the sources of exposure, such as industrial emissions, contaminated water sources, or contaminated soil, helps assess the potential risk to human health.

3. Bioavailability: Bioavailability refers to the portion of an element that can be absorbed by the body and is available for biological interactions. Not all elements in the environment are equally bioavailable. Factors such as chemical form, solubility, and interactions with other substances influence the extent to which elements are absorbed.

4. Biotransformation and Metabolism: Once absorbed, elements can undergo various metabolic processes in the body. Some elements may be metabolized, transformed into different chemical forms, or excreted. Biotransformation processes can affect the toxicity, distribution, and retention of elements in the body.

5. Biomagnification and Bioaccumulation: These concepts explain the increase in element concentrations as they move up the food chain (biomagnification) and the accumulation of elements in organisms over time (bioaccumulation). Predators higher in the food chain can accumulate higher levels of elements due to the cumulative effect of ingesting contaminated prey.

6. Half-Life and Clearance: The half-life of an element refers to

the time it takes for half of the absorbed amount to be eliminated from the body. Clearance is the rate at which elements are removed from the body through various routes, such as urine or feces. These concepts help determine the duration of exposure and potential long-term effects.

7. Target Organs and Health Effects: Certain elements have a predilection for specific organs or tissues, where they can exert their toxic effects. Understanding the target organs helps predict the health effects associated with exposure. For instance, lead is known to affect the nervous system, while cadmium can impact the kidneys.

8. Threshold vs. Non-Threshold Effects: Threshold effects occur when a certain level of exposure is necessary to trigger a health effect. Non-threshold effects occur even at very low exposure levels. Some toxic elements, like lead, exhibit non-threshold effects, meaning that there is no safe level of exposure.

9. Synergistic Effects: Elements can interact with each other or with other environmental factors to create synergistic effects, where the combined effect is greater than the sum of individual effects. These interactions can enhance the toxic effects of certain elements.

10. Risk Assessment and Regulation: Risk assessment involves

evaluating exposure, toxicity, and potential health effects to determine acceptable exposure limits. Regulatory agencies establish guidelines and standards for permissible levels of trace and toxic elements in environmental media, food, and drinking water to protect public health.

Human exposure to trace and toxic elements is influenced by a multitude of theories and concepts that help us understand how these elements are acquired, absorbed, distributed, metabolized, and can impact human health. These concepts are essential for assessing the risks associated with exposure and for developing effective strategies to mitigate potential health hazards.

ABOUT THE AUTHOR

Dr. Amit Singh Thakur

Dr. Amit Singh Thakur joined IPS Academy Indore on 14 th September 2015, he is having experience of more than 13 years in teaching and research. He taught Inorganic Chemistry, Pharmaceutical Chemistry, Medicinal Chemistry and Polymers to the students of B.Sc. and M.Sc. He is enrolled as a research guide DAVV Indore since August 2022. He has published more than 21 research papers, one book chapter, 3 books and 1 patent. His areas of interest are Pharmaceutical Chemistry, Inorganic Chemistry, Analytical Chemistry, Spectroscopy and Indian,Knowledge System

Made in the USA
Monee, IL
12 November 2026

cc757074-364d-4c6c-8836-1721cea57c8bR01